Golden Gate Beach Wrack

Words & Illustrations by Diane T Sands

ISBN 978-1-105-51332

Definition
Beach Wrack consists of dried or decaying seaweed and animal parts washed up with the tide and left stranded on the shore.

Scope
This guide was created in conjunction with the Farallones Marine Sanctuary Association (FMSA), for use during their school field trips to Crissy Field Beach. The plants and animals depicted here can be found on that beach, and at any other mildly sheltered beach in and around the Golden Gate. Beaches along the open ocean will have the most overlap with this guide. Beaches further inside the San Francisco Bay will have differing levels of salinity in the water, and the plants and animals found there will be ones that can tolerate more fresh water.

There are many books about the plants and animals that live in coastal waters. Some of these can be found in the bibliography.

Helpful Tips
By the time many of the intertidal plants and animals wash up on the beach, they have been tumbled mercilessly by the waves, resulting in broken bits and pieces. The strip of sand at the bottom of each plate shows how I found many of the species as I worked on the guide.

Many species in the tidal zones literally live on top of each other. When exploring the beach, piles of wrack can be made up of many different species. Poke at them; tease them apart and see if you can identify all of the critters you find.

Contents

Scope...3

Helpful Hints 3

Bivalves 6

Snails...8

Sponges, Hydroids, Tunicates and Bryozoans 10

Arthropods (other than crabs) 12

Crabs 14

Jellies 18

Echinoderms 20

Anemone 20

Vascular Plants 22

Green Algae 24

Red Algae   26

Brown Algae 30

Bibliography 34

Index 36

Bivalves

Two shells held together by a strong muscular valve surrounding a soft body, these shellfish are not fish, but mollusks. It is a rare and magical find to come across one with both shells still attached.

Saxidomus giganteus Butter Clam - Width to 15cm. Large black hinge ligament located on outside of white shell which is often stained with iron sulfide. Commercially harvested for clam chowder. Individual clams may live 20 years or more. Shells once used by Native Americans as money. Eats by filtering plankton from the sea water.
Predators: *Pycnopodia helianthoides; Evasterias troschelii; Polinices lewisii; Cancer magister;* sea otters; gulls.

Clinocardium nuttallii Heart Cockle - Usually less than 5cm wide, this clam can grow up to 14cm. Most are wider than they are tall. Profile from the side is heart-shaped. Shells usually light tan with stripes or splotches of brown. More than 30 radial ribs cause the shell edges to interlock. Found in eel grass beds.The pea crab, *Pinnixafaba*, often lives inside the shell. Eats by filtering plankton from water.
Predators: *Pycnopodia helianthoides; Pisaster brevispinus, Cancer magister,* gulls

Pododesmus cepio Pacific Jingle Shell - width to 10cm across. The thin valves of this oyster are nearly circular. The right valve has a hole near the hinge through which animal cements itself to rock. There is no other local oyster species similar to this one. Flesh is a bright orange. Common on pilings, rock, even plastic.
Eats plankton by filtering seawater.
Predators: *Evasterias troscheli; Orthasterias koehleri*

Mytilus californianus California Mussel - length to 10cm. Shell is heavy and blue-black; anterior end comes to a sharp point. Holds itself to rocks with red byssal threads. By breaking and creating new threads, this mussel can move from place to place, but very slowly. Can have isopods, pycnogonid crabs, pea crabs, or dinoflagellates living inside the shell. Eats by filtering sea water, about 2-3 liters/hour.
Predators: *Pisaster ochraceous,* sea birds, *Nucella emarginata, Ceratostoma nuttali, Roperia poulsoni*

1. *Saxidomus giganteus*

2. *Clinocardium nuttallii*

3. *Pododesmus cepio*

4. *Mytilus californianus*

Snails

Snails are molluscs (like the bivalves) with one shell instead of two. Snails move from place to place using their muscular bodies. They have ridged teeth, or radula, that they use to scrape algae off rocks or to drill into shells.

Tegula funebralis Black Turban – 1.9 - 4.4cm high. Purplish-black shell that is white and smooth near aperture. Found on rocks and in tide pools, may be partially adjusted to life out of water. Can crawl about 0.7mm/second. The hermit crab, *Pagurus samuelis*, uses empty shells for housing.
Eats fleshy, multicellular algae.
Predators: *Pisaster ochraceous* sea otters, *Cancer antennarius*

Littorina sp. Periwinkle - 0.3-4.4cm high. Shells are ovate to almost round. Lives on rocks and in tide pools. Eggs are laid in floating capsules. Hard to determine species without dissection.
Eats primarily microscopic algae, seaweed or sea grasses, but will feed on small invertebrates.
Predators: sea stars, crabs, sea anemones

Diodora aspera Rough Keyhole Limpet - 2.5-7cm long. Shell moderately elevated with a broad, oval hole slightly off center. White or cream with darker radial stripes. Defends itself by extending mantle folds over body and shell, giving no place for sea star predator tube feet to latch on. Also, the commensal polychete worm will bite the sea star's tube feet and drive it away.
Eats omnivorously but seems to prefer encrusting bryozoans.
Predators: sea stars, including *Pisaster*.

Lottia pelta Shield Limpet - to 40mm in length. Dark, blotchy in color, often checkered with white. Inside of shell bluish white with a dark apical blotch. "Runs" away from predators by lifting shell and crawling at top speed.
Eats red & brown algae.
Predators: sea stars, crabs and shore birds.

Mopalia muscosa Mossy Chiton - 92mm long. Oval, flat. Bony shell plates are surrounded by a girdle with stiff mossy hairs. Can be dull brown, dark olive, or gray in color. The shell plates or valves are the same color as girdle or paler with flattened ridge down midline. The valves are often overgrown with algae, barnacles, or tube worms. Found on rocks; when knocked off will roll into a ball. Valve plates are usually the only things that wash ashore.
Eats red or green algae.
Predators: Glaucous-winged Gulls; Black Oystercatchers.

1. Tegula funebralis

2. Littorina sp.

3. Diodora aspera

4. Lottia pelta

5. Mopalia muscosa

Sponges, Hydroids, Tunicates, and Bryozoans

While many of these look like plants, they are all animals living in colonies. Tunicates are chordates, the only animals in this guide to have a (primitive) spinal cord.

Halichondria bowerbanki Yellow Sponge – Forms a mass to 3cm thick and branches up to 12cm high. Tan or yellowish. Found on rocks or other animals. Tolerant of muddy conditions. A non-native, it was introduced with oysters to the Bay region
Eats plankton filtered from sea water.
Predators: sea urchins.

Aglaophenia struthionides Ostrich Plume Hydroid - to 12cm high. Beige to brown feather-like branches. Occurs in rock crevices. The skeleton shrimp, *Caprella* sp often found clinging to it.
Eats plankton filtered from sea water.
Predators nudibranchs, such as *Flabellina trilineata*

Bugula neritina Branching Bryozoan - 1mm high filter feeding individuals in colonies of hundreds. Red or yellow-purple in palmate clusters. Found on pilings or wood, never on rock.
Eats plankton filtered from sea water.
Predators: nudibranchs

Botryllus sp. Compound Tunicate - 5mm across. A bright, almost artificial orange jelly-like substance. *B. schlosseri* has regularly placed 'flowers' on surface. *B. violaceous* has irregular flower patterns. Flower patterns are easily seen with a dissecting microscope. A non-native chordate.
Eats plankton filtered from seawater.
Predators: flatworms, crustaceans, and gastropods.

Watersipora subtorquata Colonial Bryozoan - individuals 0.3-1.5mm long. Entire colony usually bright orange to brown-black when alive, large colonies have lobes and frills. Non-native.
Eats by filtering plankton.
Predators: limpets and chitons.

Haliclonia permollis Purple Sponge - 4.1cm in height. color ranges from pink to lavender to purple. A thin encrusting sponge with volcano-like waster oscula.
Eats by filtering plankton from sea water.
Predators: nudibranchs.

1. Halichondria bowerbanki

2. Aglaophenia struthionides

3. Bugula neritina

4. Botryllus sp.

5. Watersipora subtorquata

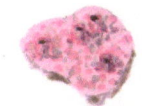

6. Haliclona permollis

Arthropods (other than crabs)

One of the cool things about arthropods is that they wear their skeleton on the outside of their body. This is good news for the beachcomber who can regularly find bits of bone along the wrack line.

Ligia occidentalis Rock Louse - to 3cm long. A variable grey-green with long appendages at end of tail. Most active at night or during dark days. Breathes air; cannot live under water.
Eats microscopic algae
Predators: crabs and shore birds

Ligia pallasii Rock Louse - to 3.5cm long. This usually grey-brown isopod has overlapping plates and breathes air. Found among rocks above high tide line as it cannot live underwater.
Eats decaying algae.
Predators: crabs and shore birds

Idotea sp. Seaweed Isopod - to 4cm long. Usually brown or green to match seaweed habitat. Breathes underwater and is a good swimmer.
eats: diatoms, algae
Predators: intertidal fishes.

Megalorchestia californiana Beach Hopper - 2.8cm body length - 6cm with antennae. White or ivory-colored with orange antennae. Spends daylight hours buried in moist sand.
Eats decaying algae
Predators: shore birds; staphylinid beetles; raccoons.

Melita nitida Common Scud – 7-9mm long. Dark greenish to slate in color. Not native to the Bay Area.
Eats: copepods and other meiofauna
Predators: fishes

Balanus improvisus Bay Barnacle- 6mm high; 13mm wide. Found on rock pilings, oysters and other hard shelled animals. Can tolerate fresh water for at least part of the year. The shell grows faster than the body.
Eats plankton filtered from water with their feet.
Predators: sea stars, crabs, goby fish.

Balanus glandula Acorn Barnacle - shell 2cm or less wide. Habitat varies from bays to rocky shores. Can live for 10 years.
Eats plankton filtered from water with their feet.
Predators: oysterdrill snails; ribbed limpets; seastars; goldeneye ducks; gulls; nemertean worm; barnacles; nudibranchs.

1. Ligia occidentalis

2. Ligia pallasii

3. Idotea sp.

4. Megalorchestia californiana

5. Melita nitida

6. Balanus glandula

7. Balanus improvisus

Crabs

Crabs are arthropods with ten legs - eight for walking and two chelipads, or foreclaws. The sand crab and the porcelain crab with eight legs each are not true crabs, though they are arthropods.

Hemigrapsus nudus Purple Shore Crab - 4-5.5cm across. Normally dark purple, but they can be red-brown or olive green. Claws have spots & white tips. Legs are hairless. Three "teeth" on edge of carapace to side of each eye. Common under rocks in high tide area. Nocturnal. Females will carry up to 36,000 eggs.
Eats: diatoms, small *Ulva* scraped from rocks.
Predators: gulls, white winged scoters, *Anthropleura* and sculpins.

Pachygrapsus crassipes Striped Shore Crab - to 4.8cm wide. Has transverse green stripes on red, purple or green. Two teeth on carapace. Large claws are red-purple above, white below. Can be found in crevices, under rocks and in clay burrows and is quite active during daytime. Is more inclined to pinch than most other species. Males larger than females. Can stay out of water for up to 70 hours.
Eats algae, diatoms, and occasionally, dead animals or limpets.
Predators: gulls, rats. raccoons, humans, sea anemones, fish.

Emerita analoga Sand Crab - not a true crab. 3.5 cm in length. A smooth egg-shaped body, flattened legs, no claws, gray with long feathery antennae tucked under the body. Is found only on wave-swept beaches where it spends most of its time buried in the sand. *Emerita* moves backward only, via the tail kept curled under the body.
Eats by filter-feeding on algae
Predators: fish, birds, sea otters.

Pachycheles rudis Thickclaw Porcelain Crab - not a true crab. Grows to 1.9cm. Mottled mix of grey, brown or white. Front claws unequal in size; the larger claw may be on either the left or the right. Found under rocks, in kelp holdfasts or hidden in barnacle shells.
Eats by filtering-feeding on plankton
Predators: fish, sea stars, drilling snails.

1. Hemigrapsus nudus

2. Pachygrapsus crassipes

3. Emerita analoga

4. Pachycheles rudis

Crabs

Romaleon (Cancer) antennarius Brown Rock Crab - 15-17 cm wide. Deep red or brown, but can be orange or grey; conspicuous red spotting on underside of body. Walking legs are almost always hairy. Found in kelp beds. Have been known to capture hermit crabs by standing over them, utilizing their legs like cage bars, The hermits can then be eaten at leisure.
Eats: bivalves, snails, echinoderms, crustaceans
Predators: sea otters, sharks, octopi, sea bass, humans

Cancer productus Red Rock Crab - to 20 cm wide; wider than long. Nearly smooth, carapace "teeth" broad rounded. Adults are uniformly red; juveniles highly varied. Fore claws have black tips. Underside white to cream with no markings. Mating occurs when females are soft-shelled after a molt. Male will protect female until her shell hardens.
Eats: clams, snails, mussels, barnacles, smaller crabs, sea cucumbers
Predators: sand bass, sculpins, gulls.

Pugettia producta Northern Kelp Crab - 7.8-9cm wide. Kelp brown or dark red in color; color is food dependant. Underside yellow or scarlet. Males have larger chelipeds (fore claws) than females. Abundant in kelp canopy or wharf pilings. Active nocturnally.
Eats primarily algae; also barnacles, mussels, hydroids, and bryozoans
Predators: sea otters sculpins, gulls, cabezon, *Velella* (will eat zoeae).

Metacarcinus (Cancer) magister Dungeness Crab - 19-23cm wide; carapace widest at 10th tooth. The claws have many serrated bumbs, white tips. Found on sandy bottoms and in eel grass beds. The largest edible crab, it accounts for 99% of all crabs taken for commercial reasons. Females can store sperm from one season to use the next. The larvae are immune to the sting of jellies and often catch a ride towards shore by clinging to tentacles, especially those of *Velella velella*.
Eats primarily clams, fish, and crabs, as well as sea stars, squid, snails, and worms.
Predators: humans, octopi, sculpins, halibut, and eels.

1. Cancer antennarius

2. Cancer productus

3. Pugettia producta

4. Cancer magister

Jellies

Bags of water with nervous systems, jellies are not fish. They aren't even vertebrates, so how can they be fish? Stop calling them that.

Polyorchis haplus Bell Jelly - bell 4-6 cm high with up to 30 tentacles. Has a long tubular stomach that runs the length of bell. Near the base of the tentacles is a ring of red, light sensitive ocelli. They inhabit the quiet bottom layer of bays or eel grass beds.
Eats: benthic crustaceans and zooplankton
Predators: fish, nudibranchs, snails and sea stars.

Aurelia aurita Moon Jelly - between 5-40cm wide. Found in both the Atlantic and the Pacific oceans, near the coast. Swims horizontally by pulsations of the bell. Decreased salinity causes a decrease in the bell arc, which is why many beached specimens are so flat.
Eats: zooplankton, most of which gets caught on the mucous-y surface of the animal and passed to the stomach by flagellar currents.
Predators: ocean sunfish (*Mola mola*), sea birds.

Pleurobrachia bachei Sea Goose Berry – spherical ctenophore to 15mm with eight rows of combs on edge of sphere. Has one pair of branched tentacles which do not sting. One of the dominant predators in CA waters.
Eats: zooplankton, especially fish larvae, small crustaceans and eggs.
Predators: comb jelly (*Beroe*), salmon and other fish.

Velella velella By-the-Wind-Sailor – about 6cm or less in width. Consists of a transparent, blue/purple, flat oval float with an erect, triangular sail perpendicular to the body (diagonal to body's long axis). Remains afloat on surface of Pacific Ocean most of its life, pushed by the prevailing winds. They wash up on CA shores in late spring -early summer. The megalopae stage of Dungeness crabs, immune to jelly stings, will grab on to outer tentacles and catch a ride towards shore, dropping to the sand where they molt into adults. Chances are if you see a lot of Velella wash up, it will be a good Dungeness crab season later that fall.
Eats: surface zooplankton, fish eggs.
Predator: violet slug, anemones, sun fish (*Mola mola*).

1. Polyorchis haplus

2. Aurelia aurelia

3. Pleurobrachia bachei

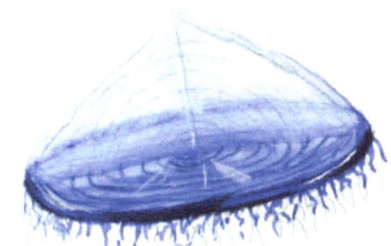

4. Velella velella

Echinoderms

Sea stars aren't fish either. They are voracious predators. When they trap their prey, they push their stomach out of their body. Stomach juices turn the prey to liquid that the star then slurps up. Yum!

Pisaster brevispinus Pink Sea Star – up to 60cm in diameter, thickest in the center. As the name implies, it is quite pink with small white spines along each radial arm. More common in bays than on the open coast. Does not survive long out of water. Will dig into sandy bottom to reach clams, a process that can take 2-3 days.
Eats: cockles, clams, snails, barnacles, sand dollars, small Dungeness crabs, also dead fish and squid.
Predators: few, possibly sea otters & gulls.

Pisaster ochraceous Ochre Star – to 25cm across. Usually has 5 rays and ranges in color from yellow to orange to brown and purple. This one is more tolerant of air than other species, able to spend more than 8 hours out of the water. That one you find on the beach might still be alive! Take a good look, and then return it to the ocean.
Eats: mussels, chitons, limpets, snails, barnacles, small crabs
Predators: Sea otters, gulls

Anemone

While there are many species of anemones along the California coast, the sandy nature of Crissy Field Beach with its absence of tidal rocks or pilings seems to keep all but the aggregating anemone from regular, predictable view.

Anthropleura elegantissima Aggregating anemone – to 6cm high and 8cm wide. Cylindrical with green tentacles tipped with pink, lavender or blue. This species prefers to live in semi-protected spots, on pilings or half buried in the sand.
Eats: copepods, isopods and other small animals grabbed by tentacles.
Predators: nudibranch *Aeolida papillosa*; snail *Epitoniium tinctum*, sea stars.

1. *Pisaster brevispinus*

2. *Pisaster ochraceous*

3. *Anthropleura elegantissima*

Vascular Plants

Vascular plants, like the algae that follow, create their own food through the process of photosynthesis. Unlike the algae they have a separate root and shoot system. Most vascular plants cannot tolerate salt water. The following are the local exceptions that get washed ashore.

Phyllospadix scouleri Surf Grass – Rounded blades less than 4mm wide. Bright green in color. They root to solid rock where the hold-fast forms a habitat for a multitude of creatures.

Zostera marina Eel Grass - flat blades usually wider than 4mm; plant up to 3m long. Roots form a hold in mud or sand. The leaves were once used as mattress stuffing and house insulation. Sea urchins, snails, crabs and a host of other critters live in eel grass beds.

cross section

cross section

1. Phyllospadix scouleri

2. Zostera marina

1.

2

Algae

Algae are classified based on the dominant pigment used in photosynthesis. The blue-green algae are mostly unicellular and are commonly called plankton, or phytoplankton. While they do wash up on the shore, they are too tiny to identify without a microscope, so they are not pictured here. The larger algae are depicted on the remaining pages and are divided into green, red and brown categories. The color often changes as it dries or decays, making identification difficult. Algae are eaten by many intertidal animals; therefore only species-specific predators will be mentioned.

Green Algae

Bryopsis corticulans Green Sea Fern - up to 15 cm long. Black-green out of the water, but bright green when alive and submerged in water. Is found on rocks, shells and wood. Each feathery arm is essentially one giant cell.

Codium fragile Sea Staghorn - up to 40 cm long. Spongy in texture; dark green to black; dichotomously branched. It will attach to any hard surface, including mussels or other shellfish. It is eaten fresh in the Philippines and Hawaii; dried in tea in Korea.

Ulvae lactuca Sea Lettuce - flat blades 2 cells thick. usually less than 30 cm long, but can grow to 1 m. pale to emerald green. common on rock, wood and other algae. When it washes up in the wrack line, it looks like wet tissue paper. Widely used for food: dried, toasted or fresh in salads.

Ulva intestinalis Green String Lettuce - thallus consists of a hollow tube up to 20cm long. In cross section, the walls are 1 cell layer thick. This species is more tolerant of fresh water and can grow in dense mats. When eaten, it is usually dried and deep fried.

Bryopsis corticulans

2. *Codium fragile*

3. *Ulva lactuca*

4. *Ulva intestinalis*

Red Algae

Calliarthron tuberculosum Bead Coral - to 25cm long. Forming dense brilliant pink turf in lower intertidal region. The sections are flat, heavily branched, with crustose holdfasts. Commonly found along the bottom of many kelp forests.

Halosaccion glandiforme Sea Sac - up to 15cm long & 2-3 cm in diameter, made up of thin-walled elongated sausage sacs. Sacs are yellow-green in sunny locations to red-purple in the shade. Usually anchored to rock, they grow in dense clusters near mussels and barnacles. Squeezing them will cause internal water to spurt through the pores.

Corallina vancouveriensis Coral Seaweed – up to 10 cm high; bright pink; forms dense mats. The branches have calcified pores for the release of reproductive cells at tips of branches. Competes for space with *Halichondria* & *Anthropleura elegantisima* and is a favorite food of the black katy chiton, *Katharina tunicata.*

Gracillaria sjoestedii Red Spaghetti - up to 1.5m tall. Typically brown to deep red; cylindrical. Is edible when blanched, pickled or fresh; called ogo in Japan, Lima Lou in Hawaii. It is also cultivated for agar production in Asia, S. America & Africa.

1. *Calliarthron tuburculosum*

2. *Halosaccion glandiforme*

3. *Corallina vancouveriensis*

4. *Gracillaria sjoestedii*

Red Algae

Chondracanthus exasperatus Turkish Towel - to 50cm tall & 18cm wide. Brick-red to purple flashy blades that are iridescent when wet; blunt at tip. Found in the low intertidal zone and on exposed rock. Its cells contain a seaweed gum called carageenan, which is found in pasta, pet food, toothpaste, ice cream and other food stuffs.

Mastocarpus papillatus Turkish Washcloth - usually less than 15cm long. Has upright red to purplish blades covered with knobby lumps or papillae, and a soft black crust (which does not usually wash up). This species is hard to tell apart from *Mastocarpus jardinii.*

Porphyra perforata Nori – Can grow to 1m in length, but is only 2 cells thick. Pale green to red to purple in color; black when dried. In the wild, it will grow on other algae, shells, and anything that does not move. It is widely cultivated in Japan and is used worldwide in sushi, soups, and other culinary delights.

1. Chondracanthus exasperatus

2. Mastocarpus papillatus

3. Porphyra perforata

Brown Algae

Nerecystis leukeana Bull Whip Kelp - up to 36m long. This is an annual algae composed of a long hollow stipe and an air bulb that supports numerous long blades (each to 4m long). It can grow 17cm per day! It grows in huge stands, anchored to rocks via holdfast. These holdfasts are host to many other algae, invertebrates, fish, etc. This algae is eaten pickled, used as fertilizer and used for feeding farm-raised abalone. It also contains an acid called algin, which is used in food production, textile screen-printing, paper manufacturing, cosmetics and a host of other ways.

Macrocystis pyrifera Giant Kelp - to 45m long. It forms vast forests which are a crucial habitat for many critters. It is the fastest growing algae - up to 0.6m per day! It is harvested as a source of algin, which is a binding agent used in food, cosmetics, ice cream, toothpaste, cereals, cake mixes and explosives. It is a favorite food of the purple sea urchin, *Strongylocentrotus purpuratus*.

Pelvetopsis limitata Dwarf Rockweed - 8-10cm tall. Olive green to light tan; inflated, warty receptacles on the tips of mature branches. Branches are flattened with NO midrib. Grows on rocky shores near the high tide line.

Fucus gardneri Rockweed - to 40cm. The tips of the branches are often inflated and a lighter green than the rest. Has a very distinct midrib on the flattened, dichotomously branched blades. This is the most common seaweed in the northern hemisphere, abundant on rocks and even mussels. This species is edible and is considered best when fresh or pickled. Individuals can live up to four years.

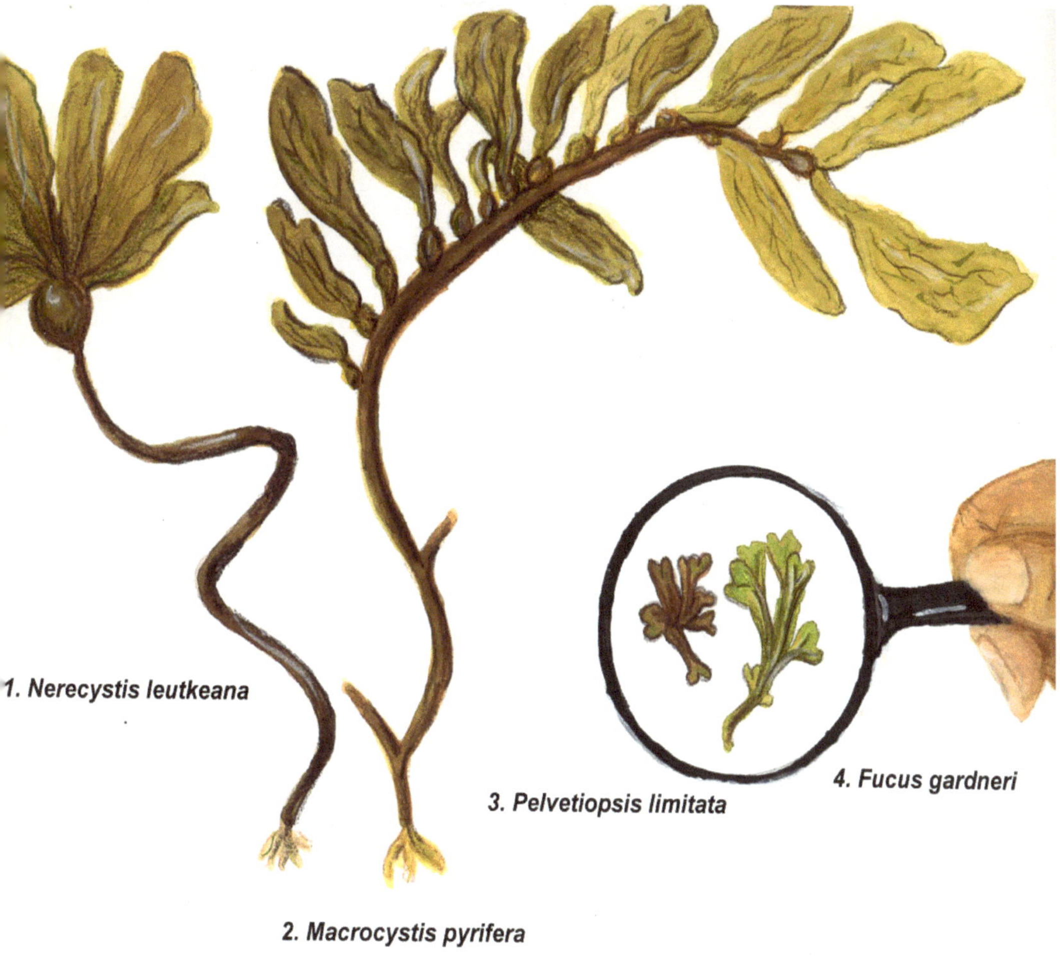
1. Nerecystis leutkeana
2. Macrocystis pyrifera
3. Pelvetiopsis limitata
4. Fucus gardneri

1.
2.
3.
4.

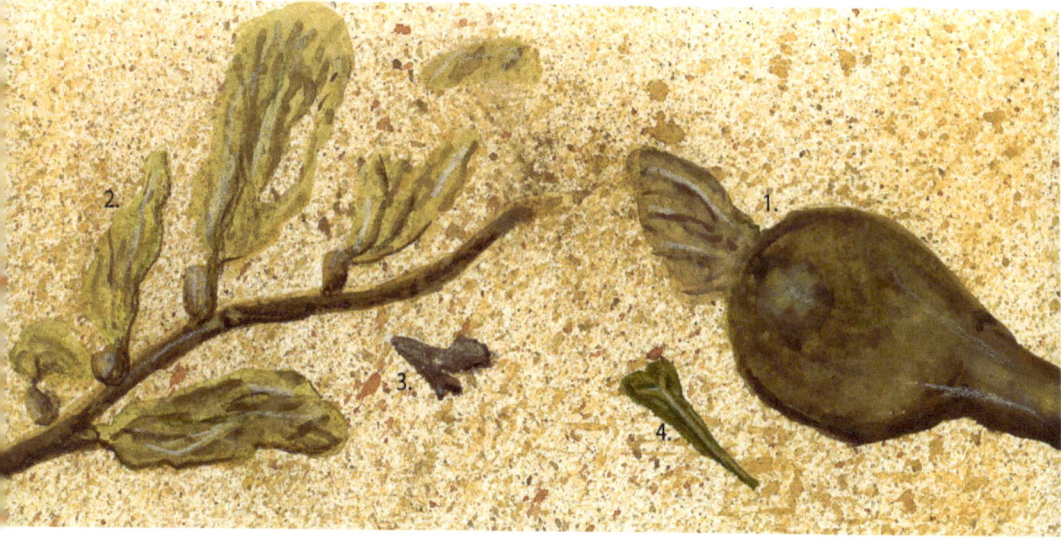

Brown Algae

Laminaria sinclarii Kombu - 5cm wide cylindrical stipe with single smooth blade up to 2m long. A fairly rubbery annual, it is one of the most readily recognized genera on California beaches. It was once harvested extensively for potash, iodine & mannitol. Is used as a soup stock flavor.

Costaria costata Five-Rib Kelp - up to 2m long. It has five midribs running the full length of the blade - three on one side, two on the other. This annual algae is yellow to a dark chocolate brown. The blades are skinny and ragged where the surf is rough, growing wider in more sheltered environments.

Egregia menziesii Feather Boa Kelp - up to 10 m long. This species can have 6-25 branches with the terminal blade the longest. Olive green to chocolate brown stipe with golden brown blades, especially near the terminal end. It grows on rocks and likes a highly saline environment. It is used as a fertilizer and as a source of alginic acid. Preferred home of the kelp crab, *Pugettia producta,* and the limpet, *Notoacmaea insessa.*

Alaria marginata Winged Kelp - to 3m long. Olive green to rich brown in color. The hollow stipe is cylindrical near the small holdfast and flattens as it reaches blade terminus. Both the blade and midrib are often used in salad; can be eaten fresh, dried or cooked.

1. *Laminaria sinclairii*

2. *Costaria costata*

3. *Egregia menziesii*

4. *Alaria marginata*

Bibliography:
Abbott, Isabella and George Hollenberg. (1976) Marine Algae of California. Stanford: Stanford University Press.

Biodiversity Heritage Library. Available from: http://www.biodiversitylibrary.org/

Cowles, Dave. (2005) Key to Invertebrates Found At or Near The Rosario Beach Marine Laboratory (a campus of Walla Walla University) Fidalgo Island, Anacortes, WA. http://www.wallawalla.edu/academics/departments/biology/rosario/inverts/index.html

Druehl, Louis. (2000) Pacific Seaweeds. British Columbia: Harbor Publishing.

Guiry, M.D. & Guiry, G.M. (2010). AlgaeBase. World-wide electronic publication, National University of Ireland, Galway. http://www.algaebase.org

Jensen, Gregory. (1995) Pacific Crabs and Shrimp. Monterey, CA: Sea Challengers.

Kozloff, Eugene. (1983) Seashore Life of the Northern Pacific Coast. Seattle: University of Washington Press.

Langstroth, Lovell and Libby. (2000) A Living Bay. Berkeley: University of California Press.

Love, Robin Milton. (1991) Probably More Than You Want To Know About the Fishes of the Pacific Coast. Santa Barbara, CA: Really Big Press.

McConnaughey, Bayard. (1985) The Audubon Society Nature Guides: Pacific Coast. New York: Alfred Knopf.

Mondragon, Jennifer & Jeff. (2003) Seaweeds of the Pacific Coast. Montery, CA: Sea Challengers.

Morris, Robert, Donald Abbott and Eugene Haderlie. (1980) Intertidal Invertebreates of California. Stanford: Stanford University Press.

Ricketts, Ed and Jack Calvin. (1952) Between Pacific Tides 3rd ed. Stanford: Stanford University Press.

Russo, Ron (1981) Pacific Intertidal Life. Rochester, NY: Nature Study Guild

Smith, Ralph and James Carlton eds. (1975) Light's Manual 3rd ed. Berkeley: University of California Press.

Waaland, J. Robert. (1977) Common Seaweeds of the Pacific Coast. Seattle, WA: Pacific Search Press.

Wrobel, David and Claudia Mills. (1998) Pacific Coast Pelagic Invertebrates: A guide to the common gelatinous animals. Monterey, CA: Sea Challengers.

Index

A

Acorn Barnacle. *see Balanus glandula*
Aeolida papillosa, as predator, 20
Aggregating Anemone. *see Anthropleura elegantissima*
Aglaophenia struthionoides, 10–**11**
Alaria marginata, 32–**33**
algae, 24–33
   Brown, 30–33
   as food, 8, 12, 14, 16
   Green, 24–**25**
   Red, 26–29
algin, under algae, Brown
anemones, 20
   Aggregating. *see Anthropleura elegantissima*
   as predators, 8, 14, 18
Anthropleura elegantissima, 20–**21**
arthropods, 8–**9**, 12–17
   barnacles, 12–**13**
   crabs, 14–17
   as food, 12, 16, 18, 20
   isopods, 12–**13**
   as predators, 10, 12
   rock lice, 12–**13**
   Sand Crab. *see Emerita analoga*
   sand hoppers, 12–**13**
   Thickclawed Porcelin Crab. *see Pachycheles rudis*
Aurelia aurita, 18–**19**

B

Balanus glandula, 12–**13**
Balanus improvisus, 12–**13**
barnacles. *see* arthropods
Bay Barnacle. *see Balanus improvisus*
Beach Hopper. *see Megalorchestia californiana*
Bead Coral. *see Calliarthron tuburculosum*
Beroe sp., as food, 18
birds as predators
   Black Oystercatchers, 8
   Glaucous-winged Gulls, 8

birds as predators continued
    Goldeneye Ducks, 12
    gulls, 6, 12, 14, 16, 20
    shore birds, 6, **8**, 12
    White Winged Scoters, 14
bivalves, 6–**7**
    as food, 16, 20
Black Turban Snail. *see Tegula funebralis*
Botryllus sp., 10–**11**
Branching Bryozoan. *see Bugula neritina*
Brown Rock Crab. *see Romaleon antennarius*
Bryopsis corticulans, 24–**25**
bryozoans, 10–**11**
    as food, 8, 16
Bugula neritina, 10–**11**
Bull Whip Kelp. *see Nerecystis leukeana*
Butter Clam. *see Saxidomus giganteus*
By-the-Wind Sailor. *see Velella velella*

C

California Mussel. *see Mytilus californianus*
Calliarthron tuberculosum, 26–**27**
Cancer antennarius. *see Romaleon antennarius*
Cancer magister. *see Metacarcinus magister*
Cancer productus, 16–**17**
Caprella sp., under *Aglaophenia struthionoides*
carageenan, under algae, Red
Ceratostoma nuttali, as predator, 6
chitons. *see* snails
Chondracanthus exasperatus, 28–**29**
clams. *see* bivalves
Clinocardium nuttallii, 6–**7**
Codium fragile, 24–**25**
Colonial Bryozoan. *see Watersipora subtorquata*
Common Scud. *see Melita nitida*
Compound Tunicate. *see Botryllus* sp.
Coral Seaweed. *see Corallina vancouveriensis*
Corallina vancouveriensis, 26–**27**
Costaria costata, 32–**33**
crabs, 14–17
    Brown Rock. *see Romaleon antennarius*
    Dungeness. *see Metacarcinus magister*
    as food, 16, 20

crabs continued
    Northern Kelp. *see Pugettia producta*
    as predators, 6, **8**, 12
    Purple Shore. *see Hemigrapsus nudus*
    Red Rock. *see Cancer productus*
    Striped Shore. *see Pachygrapsus crassipes*

D

diatoms, as food, 12, 14
Diodora aspera, **8–9**
Dungeness Crab. *see Metacarcinus magister*
Dwarf Rockweed. *see Pelvetopsis limitata*

E

echinoderms, 20
    as food, 16
    Ochre Star. *see Pisaster ochraceous*
    Pink Sea Star. *see P. brevispinus*
    as predators, 8, 12, 14, 18, 20
Eel Grass. *see Zostera marina*
Egregia menziesii, 32–**33**
Emerita analoga, 14–**15**
Epitonium tinctum, as predator, 20
Evasterias troschelli, as predator, 6

F

Feather Boa Kelp. *see Egregia menziesii*
fishes, as predators
    cabezon, 16
    eels, 16
    goby, 12
    halibut, 16
    intertidal fishes, 12, 14, 18
    salmon, 18
    sculpin, 14, 16
    sea bass, 16
    sharks, 16
    sunfish, *Mola mola*, 18
Five-Rib Kelp. *see Costaria costata*
Flabellina trilineata, as predator, 10
Fucus gardneri, 30–**31**

G

gastropods. *see* snails or nudibranchs
Giant Kelp. *see Macrocystis pyrifera*
Gracillaria sjoestedii, 26–**27**
Green Sea Fern. *see Bryopsis corticulans*
Green String Lettuce. *see Ulva intestinalis*

H

Halichondria bowerbanki, 10–**11**
Haliclonia permollis, 10–**11**
Halosaccion glandiforme, 26–**27**
Heart Cockle. *see Clinocardium nuttallii*
Hemigrapsus nudus, 14–**15**
hermit crab. *see Pagurus samuelis*
humans, as predators, 14, 16
hydroids, 10–**11**
    as food, 16

I

isopods. *see also* arthropods, 12

J

jellies, 18
    Bell. *see Polyorchis haplus*
    Moon. *see Aurelia aurita*
    Sea Goose Berry. *see Pleurobrachia bachei*
    By-the-Wind Sailor. *see Velella velella*

K

Katherina tunicata, as predator, 26
Kombu. *see Laminaria sinclarii*

L

Laminaria sinclarii, 32–**33**
Ligia occidentalis, 12–**13**
Ligia pallasii, 12–**13**
limpets. *see* snails
Littorina sp., **8**–**9**

M

Macrocystis pyrifera, 30–**31**
Mastocarpus jardinii, under *M. papillatus*
Mastocarpus papillatus, 28–**29**

Melita nitida, 12–**13**
Metacarcinus magister, 16–**17**
   under *Velella velella*, 18
Mola mola, as predator, 18
mollusks. *see* bivalves or snails
Moon Jelly. *see Auerlia aurita*
Mopalia muscosa, 8–**9**
Mossy Chiton. *see Mopalia muscosa*
Mytilus californianus, 6–7

N

Nerecystis leukeana, 30–**31**
non-natives, 10, 12
Nori. *see Porphyra perforata*
Northern Kelp Crab. *see Pugettia producta*
Notoacmaea incessa, under *Egregia menziesii*
Nucella emarginata, as predator, 6
nudibranchs, as predators, 10, 12, 18, 20

O

Ochre Star. *see Pisaster ochraceous*
octopi, as predators, 16
Orthasterias koehleri, as predator, 6
Ostrich Plume Hydroid. *see Aglaophenia struthionoides*

P

Pachycheles rudis, 14–**15**
Pachygrapsus crassipes, 14–**15**
Pacific Jingle Shell. *see Pododesmus cepio*
Pagurus samuelis
   under *Romaleon antennarius*, 16
   under *Tegula funebralis*, 8
Pelvetopsis limitata, 30–**31**
Periwinkle. *see Littorina* sp.
Phyllospadix scouleri, 22–**23**
Pink Sea Star. *see Pisaster brevispinus*
Pinnixafaba, under *Clinocardium nuttallii*, 6
Pisaster brevispinus, 20–**21**
   as predators, 6
Pisaster ochraceous, 20–**21**
   as predators, 6, 8
plankton, as food, 6, 10, 18
Pleurobrachia bachei, 18–**19**

Pododesmus cepio, 6–7
Polinices lewisii, as predator, 6
Polyorchis haplus, 18–**19**
Porphyra perforata, 28–**29**
Pugettia producta, 16–17, 32
Purple Shore Crab. *see Hemigrapsus nudus*
Purple Sponge. *see Haliclonia permollis*
Pycnopodia helianthoides, as predator, 6

R

raccoons, as predators, 12, 14
rats, as predators, 14
Red Rock Crab. *see Cancer productus*
Red Spaghetti. *see Gracillaria sjoestedii*
Rock Lice. *see Ligia* sp.
Rockweed. *see Fucus gardneri*
Romaleon antennarius, 16–**17**
Roperia poulsoni, as predator, 6
Rough Keyhole Limpet. *see Diodora aspera*

S

Sand Crab. *see Emerita analoga*
sand dollars, as food, 20
Saxidomus giganteus, 6–**7**
sea cucumber, as food, 16
Sea Goose Berry. *see Pleurobrachia bachei*
Sea Lettuce. *see Ulvae lactuca*
sea otter, as predator, 6, 16, 20
Sea Sac. *see Halosaccion glandiforme*
Sea Staghorn, see *Codium fragile*
sea stars. *see* echinoderms
sea urchins, as predators, 10, 30
Seaweed Isopod. *see Itodea* sp.
Shield Limpet. *see Lottia pelta*
snails, 8–**9**
    Black Turban Snail. *see Tegula funebralis*
    chitons, 8–**9**
    as food, 14, 16, 20
    limpets, 8–**9**
    Periwinkle. *see Littorina* sp.
    as predators, 10, 12, 14, 18, 20, 26

sponges, 10–**11**
    Purple. *see Haliclonia permollis*
    Yellow. *see Halichondria bowerbanki*
squid, as food, 16, 20
staphylinid beetles, as predators, 12
Striped Shore Crab. *see Pachygrapsus crassipes*
Surf Grass. *see Phyllospadix scouleri*

T

Tegula funebralis, 8–**9**
Thickclawed Porcelin Crab. *see Pachycheles rudis*
tunicate, 10–**11**
Turkish Towel. *see Chondracanthus exasperatus*
Turkish Washcloth. *see Mastocarpus papillatus*

U

Ulva intestinallis, 24–**25**
Ulvae lactuca, 24–**25**

V

vascular plants, 22–**23**
    Eel Grass. *see Zostera marina*
    Surf Grass. *see Phyllospadix scouleri*
Velella velella, 18–**19**
    under *Metacarcinus magister*, 16
    as predator, 16

W

Watersipora subtorquata, 10–**11**
Winged Kelp. *see Alaria marginata*

worms
    as food, 16
    as predators, 10, 12

Y

Yellow Sponge. *see Halichondria bowerbanki*

Z

Zostera marina, 22–**23**
    as habitat, 6, 18

Acknowledgements

Many thanks to Amy Dean and Pete Winch of FMSA for the project seed. Thank you to Justin Holl of NOAA for the consultations. Sara Taliaferro & Katura Reynolds were invaluable help with layout and execution. Becky Morin helped me chase down several references. Thank you all. Finally a huge thank you to Alice Steele and Elaine Cohen for your meticulous editing skills.

As a biology student, Diane T Sands raised insects, frogs, fruit flies and bacteria for college lab classes. She spent summers maintaining the control populations of stoneflies, rats & mice for graduate students. Sands quickly became immersed in the art of Scientific Illustration, creating works for graduate theses and exhibits. Having honed her illustrations for scientists at the California Academy of Sciences, she is an active member of the Guild of Natural Science Illustrators.

Sands is a member of ProArts, living in Oakland, CA with four cats, three toads, a tortoise and several tanks of insects. Her autobiofictionalographical comic, <u>Toast,</u> comes out as often as possible.

www.ingramcontent.com/pod-product-compliance
Lightning Source LLC
Chambersburg PA
CBHW041113180526

45172CB00001B/226